Copyright 2020 Nathan W. Tierney

All rights reserved. No part of this book may be reproduced or used in any manner without the prior written permission of the copyright owner, except for the use of brief quotations in a book review.

To request permissions, contact nathan@thefrontlines.com.

www.thefrontlines.com

Paperback: ISBN: 978-1-7360056-0-6

eBook: ISBN: 978-1-7360056-2-0

Library of Congress Number: 2020921409

First edition October 2020

Printed by Ingram Spark Publishing in the USA

75 Oaks LLC

1507 Horizon Park Blvd

Leander, Texas 78641

To those that have served our country and those lost, you are not forgotten.

Army vs. Navy?

I was stationed in Coronado California as a Rescue Swimmer and Aviation Warfare (AW) Petty Officer while in the Navy. It was a great life, but always wanted to fly helicopters rather than jump out of them. I did not have a college degree and no path to becoming a Naval Aviator. As a result of my shortcomings, I decided to apply for the Army's Warrant Officer flight program. While deployed to the Persian Gulf I received notice that I had been accepted. In January 1999, I drove to Fort Rucker Alabama for flight school, and was hit upside the head by the cultural change…I was definitely not in Coronado anymore…nor in the Navy. The Army was different, not in a bad way, but had its own unique way of doing things. Thankfully, I was blessed to have outstanding mentors throughout my military career and the experiences inspired me to draw these comics while in the service.

In total, 31 Presidents have served in the United States Armed Forces, and 24 of them served during wartime.

The UH-60 Black Hawk and AH-64 Apache helicopter can be disassembled to fit inside a C-130 Hercules airplane for transportation…only one at a time.

The Army is the oldest military service in the United States, and was officially established by the Continental Congress on June 14, 1775.

10 Types of Crappy F**king New Guys (FNG)

① THE TALKER – THIS TYPE OF FNG INCESSANTLY TALKS ABOUT HIS FAMILY, LIFE BACK HOME, PRETTY RAINBOWS, WHAT HE ATE FOR EVERY MEAL, FAVORITE MOVIE, SONGS, COLORS, FANTASIES WITH PINK UNICORNS, AND OF COURSE THE TIME HIS MOM FARTED AT THE DINNER TABLE.

AND DO YOU KNOW WHAT? THE SPRINKLES WERE RAINBOW COLOR! AND...

② THE MUTE – THIS FNG NEVER SAYS ANYTHING....EVEN IF HE'S ABOUT TO BE RAN OVER BY A TANK.

③ THE CONFUSED – DESPITE VERY CLEAR INSTRUCTIONS THIS FNG CAN'T EVEN SHOW UP FOR FORMATION ON TIME OR IN THE RIGHT UNIFORM. HIS DAZED AND CONFUSED LOOK IS ALL PART OF A SECRET FNG PLOT TO GIVE SENIOR NONCOMMISSIONED OFFICERS HIGH BLOOD PRESSURE.

OH OH, 1ST SERGEANT IS ABOUT TO BLOW HIS TOP.

GRRRR!!

4 **THE NAPOLEON** – THIS FNG THINKS HE'S IN CHARGE UPON ARRIVAL. HE BARKS ORDERS (AS IF ANYONE IS REALLY LISTENING) AND REMINDS PEOPLE THAT "I'M IN CHARGE". THIS FNG IS TYPICALLY A GENERAL'S KID AND WEARS HIS DADDY'S RANK TO ASSERT IS AUTHORITY.

MY DADDY IS A FOUR STAR GENERAL SO THAT MAKES ME ONE TOO!

5 **THE MOMMA'S BOY** – THIS FNG IS PARALYZED DURING ANY MILITARY TRAINING AND MAKES FREQUENT CALLS TO HIS MOTHER TO CHECK IF IT'S OK TO SHOOT A WEAPON, RUCK-MARCH, GET UP EARLY ETC. ETC. IF REPRIMANDED THIS FNG WILL CURL UP INTO THE FETAL POSITION SUCKING HIS THUMB WHILE IMAGINING THE WARMTH OF HIS MOTHER'S BOSOM.

MOMMY

WHO CAN I F*CK OVER TODAY?

6 **THE BLUE FALCON** – THIS FNG SPECIALIZES IN THROWING OTHERS UNDER THE BUS AND SCREWING PEOPLE OVER. BEWARE AND AVOID THIS BACK STABBING BASTARD!

7) THE BULLSHITTER –
THIS FNG IS SO FULL OF SHIT THAT HE OR SHE SPEWS PHYSICAL DIARRHEA FROM THEIR MOUTHS WHEN THEY BEGIN TO BS. THE FOUL STENCH OF THEIR LIQUID SHIT MIXED WITH KERNELS OF CORN IN ADDITION TO THEIR EXAGGERATIONS AND LIES MAKES A BULLSHITTER UNWANTED IN ANY UNIT.

I WOULD HAVE MADE IT THROUGH BUDS, BUT I WAS OVER QUALIFIED...

8) THE APOLOGIZER –
THIS FNG IS A PATHOLOGICAL APOLOGIZER THAT DOES NOT LIVE IN REALITY. NO MATTER THE CAUSE OR WHO IS RIGHT OR WRONG HE OR SHE WILL APOLOGIZE. AS EXAMPLE, IF HE OR SHE IS IN THE ARMY THEY WILL APOLOGIZE FOR THE CRAPPY FOOD ON NAVY SHIPS.

I'M SO SORRY. SO SORRY FOR YOUR GRAY SHIPS.

9) THE QUICK DRAW –
A FNG WHO KNOWS JUST ENOUGH TO BE REALLY DANGEROUS TO EVERYONE ELSE, BUT NOT ENOUGH TO ACTUALLY BE GOOD AT THEIR JOB.

I ALREADY KNOW ABOUT FIREARMS AND EXPLOSIVES, I DON'T NEED THIS TRAINING. WE USED TO SHOOT AT TANNERITE REEFERS ALL THE TIME BACK ON THE FARM. WATCH, I'LL SHOW YOU WITH THIS C-4.

10) THE APPEASER –
THIS FNG ALWAYS ACCEDES TO DEMANDS EVEN IF IT MAKES NO SENSE. HE OR SHE TRIES TO MAKE EVERYTHING CALM BY GIVING CONCESSIONS AT THE EXPENSE OF PRINCIPLES. AS WINSTON CHURCHILL SAID, "AN APPEASER IS ONE WHO FEEDS A CROCODILE – HOPING IT WILL EAT HIM LAST."

Stay Frosty is a phrase that means to keep one's emotions under control and to stay alert and on one's toes.

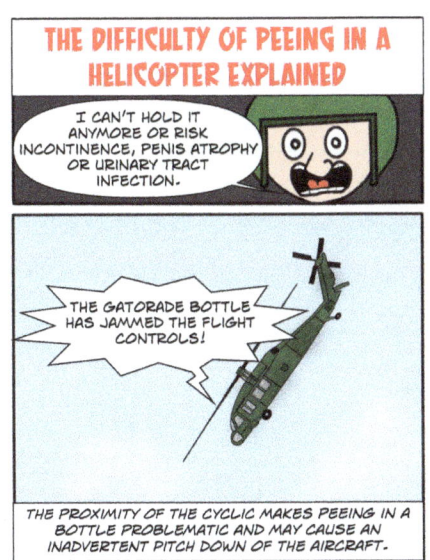

Holding in large amounts of urine for an extended period of time exposes your body to harmful bacteria, which can increase your chances of a Urinary Tract Infection or atrophy.

The term **Oscar Mike** comes from military radio jargon used on the front lines. It translates to **On the Move.**

Twenty states and the District of Columbia allow cousins to marry; six states permit first-cousin marriage only under certain circumstances.

Germany in the mid-1930s created the first night vision goggles, and were used by German tanks and infantry during World War II.

A 1998 conversation drawn while on deployment in the Pacific Ocean. Turns out Indonesia is 27th in the world for oil reserves. Shows how little I knew back then.

True radio communications of an Apache escorting a pair of Black Hawks to Kirkuk, Iraq.

Fart lighting, also known as pyroflatulence, is possible because flatulence is partly composed of flammable gases like methane and hydrogen so it can be set on fire.

When I first became a new Warrant Officer there were a couple of lively "old school" senior Warrants that had a knack for sarcasm. They were always professional with the commissioned officers, to include new Lieutenants, but the senior Warrants had a sense of humor that was priceless. In rare occasions we had some fresh Lieutenants (and a Captain) that did not understand the role of Warrant Officers as the technical and tactical subject matter experts. Their advice was ignored to the peril of the unit. Poor leadership definitely caused a few problems including negative impact on morale. Fortunately, through humor and sarcastic banter we got through those tough times. Many comics were drawn during this period and it was through awesome mentors that I could always find something to laugh at - Positive Mental Attitude (PMA is <u>not</u> PMS).

This was the original drawing I came up with, but later changed it to a one panel comic on the next page. Still undecided on which one I like more.

Officially, the birth date of the Army Warrant Officer Corps is July 1918, when Congress established the Army Mine Planter Service as part of the Coast Artillery.

Did you know that a Chinook helicopter can carry 11 tons and more than 45 troops? It also has a version, the MH-47, that can aerial refuel in flight.

Apache helicopters are featured in 30 movies or documentaries, but never King Kong. Firebirds sucks!

Original sketch after finding a Private wandering around the flight line searching for a bucket of rocket fuel. For those that are wondering there is no such thing.

Boot camp as we know it today began in the early 1900s. As example, the first Marine Corps recruit depot opened at Port Royal, South Carolina on June 1st, 1911.

Dwarf-tossing also called midget-tossing, originated in Australia in the 1980s and came to the U.S. shortly thereafter. In 1989, Florida made it illegal, and is just one of two States to ban the tossing of little people.

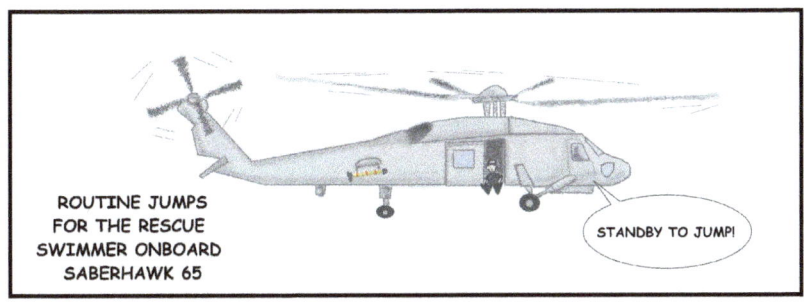

Rescue Swimmer School has more than a 50% attrition rate and are members of the Navy Special Operations community.

More than 80% of the population is afraid of flying. Acrophobia is defined as a fear of heights.

The SR-71 Blackbird is the fastest air-breathing aircraft, reaching 2,193.13 mph in July 1976. The average liberty a sailor gets in a foreign port is 48 hours before heading back out to sea again. Not a lot of time to drink, fall in love and see the cultural sights.

Daily, we use 8-9 sheets of toilet paper every time we go to the bathroom – that's 57 sheets every day! The average roll has 150 sheets so that's 11 rolls per month!

Mitch from Baywatch will never be as cool as a Navy Rescue Swimmer. Who would Pam really want to save her?

I could not decide what direction to take this comic and felt that the prior page's "So Others May Live" comic was too busy. Previously, there was too much dialogue and things going on for a one panel comic. In this version, I tried to simplify it and streamline the humor. I am not sure which is better? This is one of those occasions where I have second guessed myself on a comic and label this as a "throw away" because of its imperfections.

Below is what I call also call a "throw away" comic. I drew it one morning when I realized that we had ran out of coffee. Thank goodness there is a Starbucks is half a mile away!

Helicopters flying at low altitude can cause a vertical down wash of air known as 'rotor wash' in excess of 30mph.

The divorce rate of AH-64 Apache helicopter pilots was approximately 89% in the late 1990s, which led to the not so funny phrase amongst its pilots, "You got AIDS" (Apache Induced Divorce Syndrome).

Motrin can lead to stomach bleeding and increase your risk of fatal heart attack or stroke, especially if you use it long term or take high doses, or if you have heart disease.

Originally, one AH-64A Apache helicopter cost over $14 million, but the cost increased with each new variant to more than $35 million for the AH-64E. Interestingly, Greece paid an average of $56.25 million for the AH-64D when ordering only 12 helicopters.

Georgia Parole was an American term for shooting a British (enemy) soldier rather than take him prisoner.

According to the Economic Policy Institute, the average retirement savings of all working-age families (32 to 61 years old) is $95,776, and 15% of Americans have no retirement savings at all.

In 2020, the Secretary of the Army decided to change the Active Duty Service Obligation (ADSO) from six years to ten years following flight school to address the shortage of pilots in the Army.

The Continental Navy was formed on October 13th, 1775 and was disbanded after the war in 1785. Our country went without a navy for nine years, however, this lack of defense made us an easy target for pirates. Under the Naval Act of 1794, our country's navy was resurrected and re-named the U.S. Navy.

The Air Force was officially established in 1947 when they broke from the Army's Air Corp…the Air Force has regretted that decision ever since.

Fraternization is a violation of the Uniformed Code of Military Justice (UCMJ) and forbids personal relationships between military service members of different ranks and certain positions of authority. Sodomy and adultery is also illegal.

More than 1 in 4 (26%) adults in the United States have untreated tooth decay. The Navy provides free dental care to its sailors…even SEALS.

Over 4,000 people were rescued in the aftermath of Hurricane Katrina by military Search and Rescue forces.

AHOY is an energetic greeting to a fellow shipmate. It is like 'hello'. **ALL HANDS ON DECK** is an urgent request for help to all members of a ship's crew.

Tear gas works by irritating mucous membranes in the eyes, nose, mouth and lungs. It causes crying, sneezing, coughing, difficulty breathing, pain in the eyes, and temporary blindness...it can also cause you to vomit...trust me on puking.

The length of initial entry or Basic Combat Training (BCT) has waned over the years, ranging from as long as 17 weeks (1943) to as short as 8 weeks (1980).

The Air Force has over 70 golf courses at Air Force Installations worldwide. Air Force regulations state: 1) build the golf course 2) build the officers club 3) build the runway 4) ask Congress for more money to buy planes.

Six Warrant Officers have received the Medal of Honor for bravery during wartime.

Prior to 2009 there were seven bars at the International Security Assistant Force (ISAF) headquarters, including the sports bar Tora Bora. However, after an inability to respond to an attack General Stanley McCrystal banned alcohol from all premises.

SCIENTIFIC INSIGHT INTO HOW THESE COMICS ARE CREATED.

Most of the time my comics are inspired by real life events or pranks that I either participated in or witnessed while in the military. Many comics were originally drawn on bar napkins. However, sometimes the comics are inspired by things I have read or jokes that I have heard. As example, here is a joke that inspired a comic about General's standing around a camp fire. Based on the joke I tried to narrow it down into a one panel comic (on the next page).

Two Generals of the Army and Marines are joined by an Admiral of the Navy around a campfire off the landing zone doing shots of rye whiskey when someone calls out and asks who's got the most balls.

The Marine General goes all right and says, "Marine," over the radio, "I want you to take that beach head, clip though that barbed wire and take the entrenchments." The Marine goes, "Aye, aye, Sir" and climbs over the ship into a landing craft and sails into harm's way. He unloads and hits that beach head, breaks through the barbed wire and takes everybody in the trench except one that kills him. The Marine General goes, "That's balls!"

So, the Army General goes, "Soldier, I want you to swim ashore and take out that last man." The soldier goes, "Yes Sir!" He swims ashore and takes out the last man only to be taken down by machine gun fire.

So, the Navy Admiral says, "Sailor, bring me a cup of Joe and..." "Fuck you Sir." came across the radio. "Gentlemen," the Admiral says, "That's some balls!"

When drawing a comic, I sketch it out using shapes to outline what it will look like. Later, I add more details to the characters, color and complete the word balloons. I'm not a fan of long drawn out comic strips. Unfortunately, just because there is a good punchline it doesn't always mean I can draw it in 1-3 panels.

An AGM-114 Hellfire missile travels 955 miles per hour and an Apache helicopter can carry 16 of them.

Drug users in the United States spent approximately $100 billion annually over the past decade on illicit drugs.

Most have heard of Operation Market-Garden during World War II (thanks Sean Connery for the awesome film!), but what about the other classic and somewhat hilarious names of other major operations? Here's a few facts that inspired the comic.

Operation Gaff

This was the code name for the 1944 Allied attempt to either kill or capture Field Marshall Erwin Rommel in France. The commando attempt failed and Rommel was later implicated in the attempt to assassinate Adolph Hitler, after which Rommel committed suicide vs. face a firing squad.

Operation Beggar

In 1943, the U.S. Army Air Force scheduled a drop of weapons and ammunition to the French Resistance during the Paris uprising. The operation was cancelled at the last minute; instead food and coal were flown to the city.

Operation Aphrodite

The Allies attempted to guide remote controlled bombers to destroy hard to hit Nazi targets. An engineer and pilot took the plane off, armed the explosives in the air, and then bailed out. A separate aircraft would then use a remote control to guide the plane to the target. The thought was this process would minimize losses to bomber crews, but in fact was pretty risky. In fact, Joseph Kennedy Jr. was killed on one of these missions when his plane exploded in the air.

Operation Cartwheel

In June 1943, General Douglas MacArthur and Admiral William Halsey executed a nine-month combined operation in the Pacific Ocean. Admiral Halsey's forces attacked New Georgia Island in the Solomons at the same time as General MacArthur's forces occupied the islands of Woodlark and Kiriwina off New Guinea.

Operation Magic Carpet

This was the effort to transport U.S. servicemen back home from the European theater.

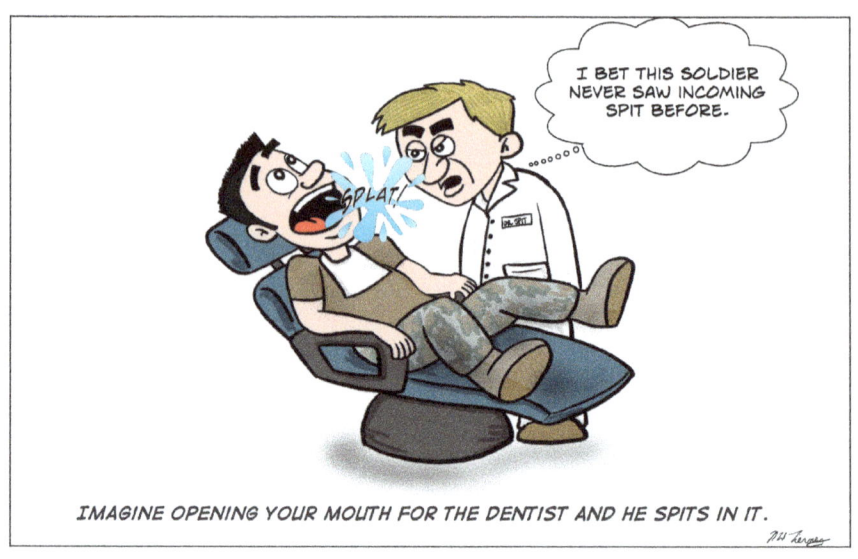

There are an estimated 75 million seals in the world, but only 2,450 are active duty SEALS.

There are 6,646 divorces per day in the U.S. (2,419,196 a year). Stealing clothes is one of the top reasons why relationships end.

Payday lenders are some of the most notorious financial predators in the U.S., and make their money by advancing people money against their paychecks. Sometimes the interest rate is 900 percent or higher! Even the mafia didn't charge rates like that!

In 2019, there were 15 fatal skydiving accidents in the U.S. out of roughly 3.3 million jumps!

There are approximately 3,862 strip clubs in the United States and industry revenue increased to $8.0 billion in 2019. Portland Oregon (yes Portland) has the most strip clubs per capita. Despite COVID-19 they opened a drive through strip club too.

As of September 10, 2020, the average weekly pay for a Hooters Girl in the United States is $650 a week. On average a Hooters Waitress makes $5 per hour, but can range from $2 to $16 per hour.

One of the safest places to be during an apocalypse is in an underground bunker or bomb shelter. The luxury Aristocrat bunker comes with a gym, sauna, swimming pool, game room, bowling alley, espresso machine and media room with theater seats for only $8.3 million!

The term **'GRUNT'** was originally a derogatory term for Army or Marine infantrymen (referencing the sounds made by men carrying heavy gear).

Benjamin Franklin sold chocolate in his print shop in Philadelphia and chocolate boosts a person's sex drive, which probably contributed to Franklin having four kids!

In the United States, 36.5% of adults are overweight and only 23 percent of Americans are meeting the federal standards for exercising. It proves comfort kills!

According to Pew Research, the United States trails other developed countries in voter turnout. Only 55.7% of Americans voted in the last Presidential election (2016).

Military uniforms are not covered by taxpayers. In boot camp, military personnel are given an allowance for their uniforms, but it comes out of their paycheck. Military members also have to pay for the medals they earn, which can range from $15 to $60 each.

"To care for him who shall have borne the battle."
- Abraham Lincoln

About the Author

Nathan Tierney is a best-selling author, artist, founder of The Frontlines (www.thefrontlines.com), Veteran of both the Navy and Army, and most importantly a father who wants to share some humor and history of life in the military through stories and comics.

Nathan began his military career as a U.S. Navy Rescue Swimmer in where he was awarded our nation's highest peacetime medal for heroism for an at sea rescue. After five years of distinguished service in the Navy, he was selected for transfer and served 14 years as a Warrant Officer in the Army, where his exceptional motivation and integrity propelled him to excellence as an Apache helicopter instructor pilot and later as a member of Special Operations. Nathan retired from the military and now lives happily ever after with his family.

Nathan is also the author of *Value Management in Healthcare* and a global speaker on the topic of value-based care. Most recently, inspired by his son's medical journey, he is leading a team of active citizens to create an application focused on transparency in healthcare through outcomes-based measurement.

www.ingramcontent.com/pod-product-compliance
Lightning Source LLC
Chambersburg PA
CBHW040108120526
44589CB00040B/2826